A rainbow is a lovely sight! When people see one, they stop to admire its colors. But just try asking some people, "What is a rainbow?" It soon becomes clear that few know the answer! Read on to discover what a rainbow really is.

A rainbow isn't something you can feel with your hands. That's because it's made of sunlight. You might think sunlight looks colorless or white when it comes through your window. But sunlight is really made up of these colors: red, orange, yellow, green, blue, indigo (ĭn´ dĭ gō), and violet.

The colors in sunlight are always there, but we only get to see them sometimes. When sunlight passes through raindrops, the light is *refracted* (rē frăc´ təd), or bent. That splits the sunlight into bands of colored light, which form a rainbow.

red
orange
yellow
green
blue
indigo
violet

If you have a three-sided piece of glass, called a prism, you can discover something exciting. When sunlight shines through a prism, it bends and splits the sunlight into the same bands of colors you see in a rainbow!

You'll also see that the colors are arranged in the same order as in a rainbow. Red is at the top, above all the others. Violet is at the bottom. The colors orange, yellow, green, blue, and indigo come between them.

When you look at a rainbow, you might not always see all seven colors. Sometimes the colors blend, and you see only three or four. How many colors do you see in the rainbow pictured above?

Sometimes you'll see one rainbow above another, which is truly a wondrous sight! A paler rainbow appears above a brighter one. The bands of color in the paler rainbow are in reverse order.

When is it a good time to discover a rainbow? Try looking for one late in the afternoon, after a rainstorm. Another good time would be early in the morning after it rains. Turn your back to the sun and look in front of you. Then, if the timing is right, you might be rewarded with a wonderful, colorful sight!